I0821845

DINGOES

by Martha London

Cody Koala

An Imprint of Pop!
popbooksonline.com

abdobooks.com
Published by Pop!, a division of ABDO, PO Box 398166, Minneapolis, Minnesota 55439.

Printed in the United States of America, North Mankato, Minnesota.

052021
092021
THIS BOOK CONTAINS RECYCLED MATERIALS

Cover Photo: Shutterstock Images
Interior Photos: Shutterstock Images, 1, 5 (bottom left), 6, 9, 13, 15, 19 (bottom right), 20; iStockphoto, 5 (top), 5 (bottom right), 10, 19 (top), 19 (bottom left); I. Bartussek/Arco Images GmbH/Alamy, 16

Editor: Aubrey Zalewski
Series Designers: Laura Graphenteen and Colleen McLaren

Library of Congress Control Number: 2020948311
Publisher's Cataloging-in-Publication Data
Names: London, Martha, author.
Title: Dingoes / by Martha London
Description: Minneapolis, Minnesota : Pop!, 2022 | Series: Desert animals | Includes online resources and index.
Identifiers: ISBN 9781532169687 (lib. bdg.) | ISBN 9781098240615 (ebook)
Subjects: LCSH: Dingo--Juvenile literature. | Wild dogs--Juvenile literature. | Dogs--Behavior--Juvenile literature. | Desert animals--Juvenile literature.
Classification: DDC 591.754--dc23

Hello! My name is

Cody Koala

Pop open this book and you'll find QR codes like this one, loaded with information, so you can learn even more!

Scan this code* and others like it while you read, or visit the website below to make this book pop.

popbooksonline.com/dingoes

*Scanning QR codes requires a web-enabled smart device with a QR code reader app and a camera.

Table of Contents

Chapter 1

Australia's Wild Dogs

Dingoes look like dogs. They have sharp teeth and long **snouts**. Most dingoes live in Australia. Many live in deserts there. They have to live near fresh water.

Watch a video here!

Dingoes live in packs. Packs are made up of approximately ten dingoes. They are a family. Packs have two dingoes that are in charge of the group.

Dingoes live in other **habitats** besides deserts. They can be found in mountains, forests, and grasslands.

Chapter 2

Desert Life

Dingoes travel long distances to find food. They hunt within their **territory**. They howl to mark their territory and to find other pack members.

Learn more here!

Dingoes have large, pointed ears. They can almost turn their heads in a complete circle. Dingoes have strong senses of hearing and sight. This helps them find **prey** in the desert.

Dingoes blend into their **habitat**. Their short fur is reddish brown or yellow brown. They often have white patches on their paws and bushy tails. Their colors match the desert. It is difficult for prey to see them.

ear
snout
tail
paws

Chapter 3

Catching Food

Dingoes are **predators**. They mostly eat rabbits and rodents such as rats. Sometimes they eat lizards or birds.

Learn more here!

Dingoes sometimes hunt in packs. Other times they hunt alone. Dingoes often look for easy things to eat, such as dead animals and garbage. Dingoes share food with their pups.

Dingoes hunt the most at night.

Chapter 4

Life in a Pack

Dingoes are **social** animals. They like to spend time with other dingoes. Dingoes mate for life. A female dingo gives birth once a year. She has four to ten pups.

Complete an activity here!

All dingoes in the pack help raise pups. Some male pups leave when they are six to eight months old. They travel on their own and make new packs. Dingoes live up to 13 years.

Dingoes create **dens** in holes left by other animals.

Making Connections

Text-to-Self

Would you like to see a dingo in real life? Why or why not?

Text-to-Text

Have you read books about dogs? How are dogs similar to dingoes? How are they different?

Text-to-World

Dingoes are predators. What are two other examples of predators?

Glossary

den – a cave or hole that shelters animals.

habitat – the area where an animal normally lives.

predator – an animal that hunts other animals for food.

prey – an animal that is hunted by other animals.

snout – the area of an animal's face that includes the nose and mouth.

social – enjoying the company of others.

territory – an area of land that is marked by an animal or pack.

Index

Online Resources

popbooksonline.com

Thanks for reading this Cody Koala book!

Scan this code* and others like it in this book, or visit the website below to make this book pop!

popbooksonline.com/dingoes

*Scanning QR codes requires a web-enabled smart device with a QR code reader app and a camera.